All About Antarctic Penguins

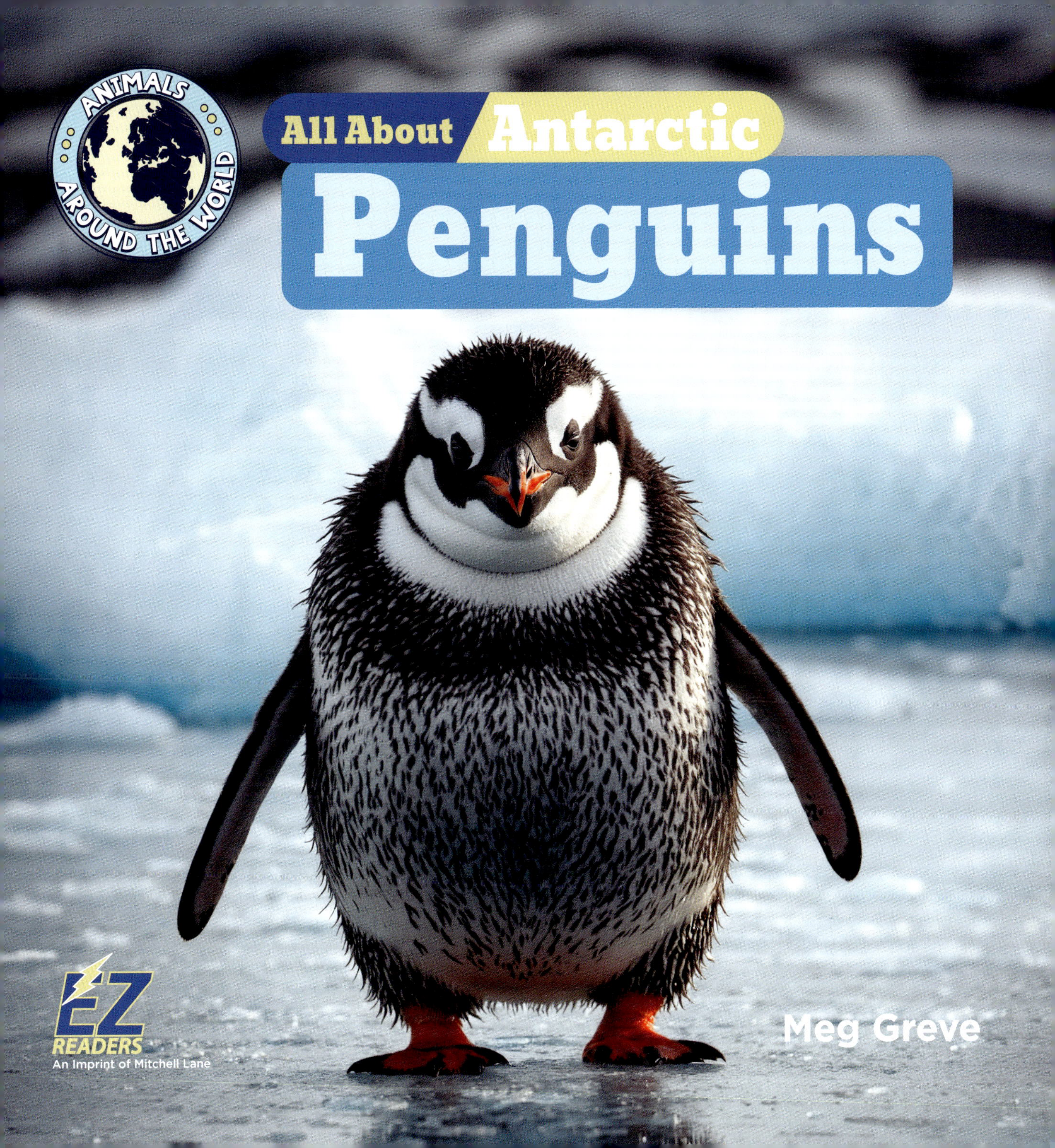

Meg Greve

EZ READERS
An Imprint of Mitchell Lane

Creating Young Nonfiction Readers

EZ Readers lets children delve into nonfiction at beginning reading levels. Young readers are introduced to new concepts, facts, ideas, and vocabulary.

Tips for Reading Nonfiction with Beginning Readers

Talk about Nonfiction

Begin by explaining that nonfiction books give us information that is true. The book will be organized around a specific topic or idea, and we may learn new facts through reading.

Look at the Parts

Most nonfiction books have helpful features. Our *EZ Readers* include color photographs and graphic aids, a table of contents, a glossary, and an index. Share the purpose of these features with your reader.

Color Photos and Graphic Aids

A lot of information can be found by "reading" photos, charts, maps, and other graphic aids found within nonfiction texts. Help your reader learn more about the different ways information can be displayed.

Table of Contents

Located at the front of the book, this list shows the big ideas within the text and the page numbers where they can be found.

Glossary

Located at the back of the book, the glossary defines key words and phrases that are related to the topic. These words and phrases can be found in the text in colored type.

Index

Located at the back of the book, an index is an alphabetical list of topics and the page numbers where they can be found.

With a little help and guidance about reading nonfiction, you can feel good about introducing a young reader to the world of *EZ Readers* nonfiction books.

Mitchell Lane
PUBLISHERS

2001 SW 31st Avenue
Hallandale, FL 33009
mitchelllanepub.com

Copyright © 2026 by Mitchell Lane Publishers. All rights reserved. No part of this book may be reproduced without written permission from the publisher. Printed and bound in the United States of America.

First Edition, 2026.

Author: Meg Greve
Designer: Rhea Magaro
Editor: Kim Thompson

Names/credits:
Title: All about Antarctic Penguins / by Meg Greve
Description: Hallandale, FL :
Mitchell Lane Publishers, [2026]

Series: Animals Around the World
Library bound ISBN: 979-8-89260-519-9
eBook ISBN: 979-8-89260-522-9

Library of Congress Control Number: 2025933804

EZ Readers is an imprint of
Mitchell Lane Publishers

PHOTO CREDITS
Shutterstock: MuhammadHanif1, cover, 1; Cathleen D, 4; pixuberant, 4; Tetris Awakening, 5; juan68; Zaruba Ondrej, fieldwork, 5; Rob Jansen, 5; Tom Volkpv, 7; Valerie VBBN, 8; AnnaHappy, 10; David Herraez Calzada, 12; slowmotiongli, 14; MZPhoto. CZ, 16; Goinyk Prtoductions, 19; Sandra Ophorst, 21; deer boy, 22;

Contents

Antarctica is covered in snow and ice. All around it, the Southern Ocean is filled with icebergs and glaciers. Not many animals live here. But penguins do! Seven kinds of penguins live in this part of the world.

Adélie Penguin

King Penguin

Chinstrap Penguin

Gentoo Penguin

Emperor Penguin

Macaroni Penguin

Rockhopper Penguin

Antarctic penguins have a thick layer of **blubber** under their skin. Their feathers are coated in oil to make them **waterproof**. The feathers are packed close together so that penguins stay warm and dry.

A penguin can be as short as a bowling pin or as tall as a baseball bat. All penguins have black and white feathers. Some, like king penguins and emperor penguins, also have yellow feathers.

Penguins are **flightless** birds. On land, they waddle and hop. They slide on their bellies over ice and snow. Their **webbed** feet help them paddle in shallow water.

Penguins are excellent swimmers. They spend almost their entire lives in the ocean. They use their strong wings to "fly" through the water.

Penguins dive deep in the ocean to find food. They hunt for fish, **krill**, and squid. Their tongues are covered in spikes. This helps them hold on to their slippery meals.

Penguins live in groups called **colonies**. Colonies are found on beaches and in rocky places. Macaroni penguin colonies can have millions of penguins. They are noisy and messy!

All Antarctic penguins lay their eggs on land or ice. They lay one or two eggs each year. Once they hatch, the chicks must wait for their adult feathers to grow in before they can swim.

Adélie and gentoo penguins build nests from pebbles. Both parents help keep the eggs warm. They chew and spit up fish for the chicks to eat.

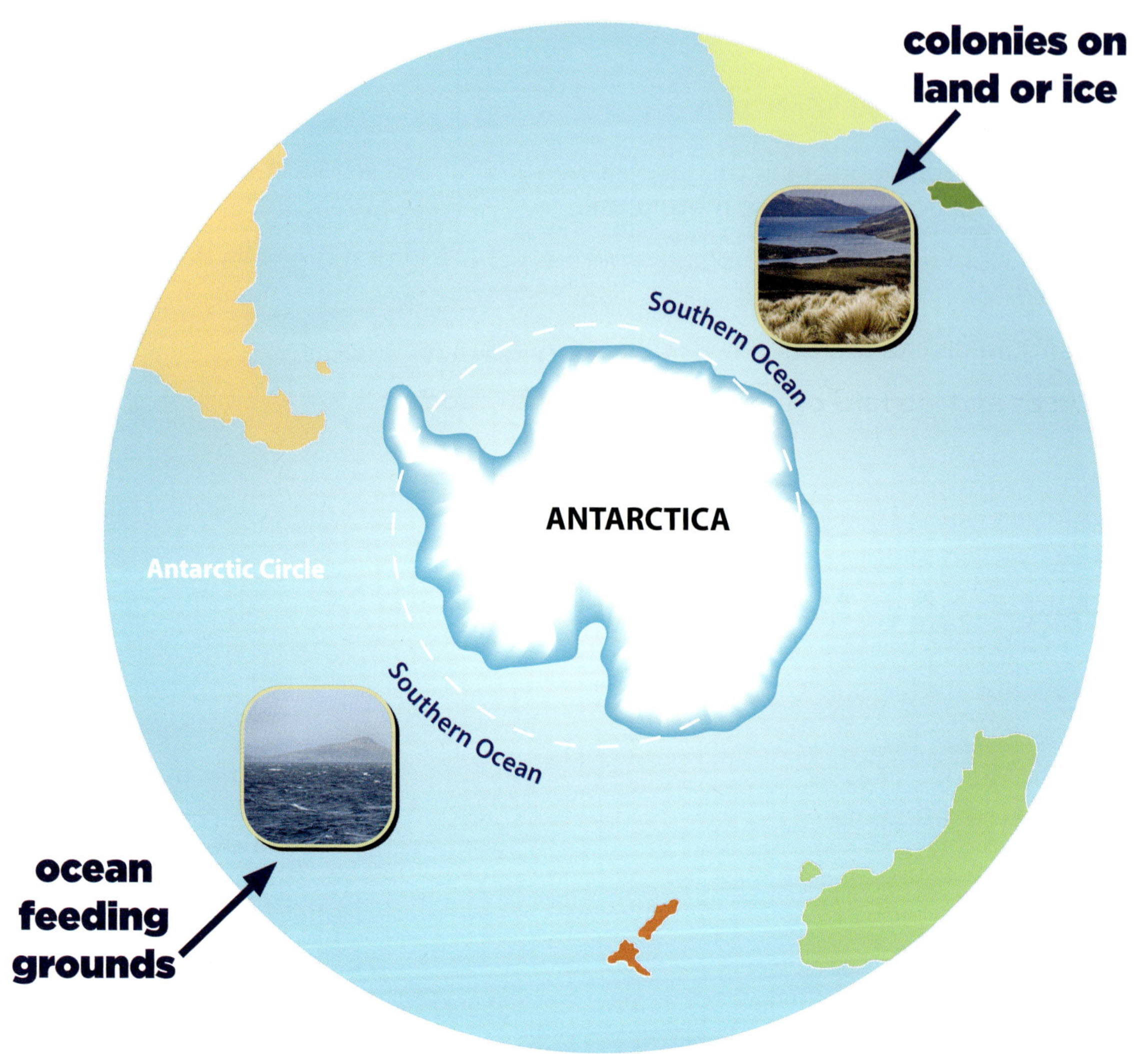

Where Do Penguins Live?

Seven kinds of penguins live in Antarctica and the Southern Ocean. They feed in the ocean, but they nest on land or ice.

Interesting Facts

- The emperor penguin is the largest penguin in the world. It can stand over four feet (one meter) tall.
- A penguin rookery is the place where eggs are laid and chicks are cared for.
- Penguin chicks are born with soft, fluffy feathers.
- Orcas and leopard seals hunt penguins for food.

Parts of a Penguin

beak or bill
A penguin's beak is sharp and pointy.

eyes
Penguins have excellent underwater eyesight to help them spot prey.

feathers
Penguin feathers are packed tightly together. This helps penguins stay warm and dry.

feet
Webbed feet help penguins swim and dive. Their feet have claws to help them grip the ice.

wings
Penguin wings are stiff like paddles. They help penguins swim very fast underwater.

Glossary

blubber (BLUHB-ur)
A layer of fat under the skin of an animal that lives in a cold place

colonies (KAH-luh-nees)
Large groups of animals

flightless (FLITE-les)
Not able to fly

krill (kril)
Small, shrimp-like creatures that are an important source of food for many ocean animals

waterproof (WAW-tur-proof)
Able to stay dry in the water

webbed (webd)
Having skin that connects the toes

Index

Further Reading

Donaldson, Julia. *Jonty Gentoo: The Adventures of a Penguin.* Scholastic Canada, 2024.

Kellett, Jenny. *The Ultimate Penguin Book for Kids.* Bellanova Books, 2022.

On the Internet

Cool Kid Facts: Interesting Penguin Facts
www.coolkidfacts.com/penguin-facts
Learn fascinating facts about these flightless birds.

National Geographic Kids: Emperor Penguin
kids.nationalgeographic.com/animals/birds/facts/emperor-penguin
Learn about the most famous penguin of all.